Bent Truth

Warning! **Don't read before bed!**

By Lawrence V. Webster USNR (Ret.)

Forward

When looking into the alien phenomenon, we sometimes forget to ask ourselves basic questions. Can we answer our own questions? If not, why? Let's ponder upon the question of why would aliens allow themselves to crash in Roswell, New Mexico in 1947? We need not use any history as it stands. Why? History on this subject is skewed at best. Accounts are useful if you do not ask the right questions, so you will have a foundation to stand on. But, what if that foundation was sending everyone on a wild goose chase to continue to distort the truth! Why must we not look to our government for answers concerning these events? For too many years, we have relied on leaders for answers they are not willing to divulge. Relying on others for what we can do ourselves presents a problem as well.

The answer to the question of did a spaceship land or crash in New Mexico will be answered in one paragraph or less, then we can move on! The government, meaning the Air Force, already answered that question for everyone. So why are we still asking it! The answer is yes! The Air Force admitted in 1947 that they have captured a flying disk in Roswell. Be careful not to stump your toes on information that is not relevant. If you can use it, use it. If not, move on! Which is precisely what I intend to do on this subject. At certain intervals in this book, I will present you with some mind-blowing scenarios. You will have to decide what you can use, or not use. Think on your own! I have purposely not added page numbers to this book, to break the simple thinking pattern we as humans tend to have. Remember where you were, and you'll come back! I would

suggest, finding the most unforgettable book mark you can find. You're going to need it.

Chapter One

Why give us what we can't give ourselves

If you are ready for a fascinating journey into the known or expected, let us begin now. Wasting time is not one of my strong suits. I am however cautious! Because when a human decides to think on their own without being told what to do, is in fact, mind-blowing! We all have the inherent right of life, liberty, and the pursuit of happiness! We were all given a choice to do, or not do! To obey God, or not to obey! Each choice you make has its own set of circumstances and or consequences! Mind-blowing right? So, armed with this and a few more shields, and swords, I will attempt to provide you with, I expect that you will use it wisely, or not. Once again, you have a choice! A choice to use what God has given us to think accordingly or…not! That's already too much information. Man has inherent flaws that go way back to the story of Adam and Eve! Do we not? These flaws are what make us unique. Sometimes, these very flaws can cause us to behave exactly the way someone or something else expects us to behave. Understanding our flaws can help us to advance and move ahead at a rapid pace. However, very few seem to take that path.

We may speak of some great people in our past that seemed to have jumped the very frame of time they were living in, to bring us a new sense of self. Ideas that were meant for us to advance into the next millennium, to reach the very lights in the skies that we tend to stare at in wonder at each and every night. You too, are a magnificent creation, that when put to the test, will use those abilities that we all have to leap right out of the time frame that we are living in. As to take us to places we can only imagine.

If you ask me if we possess the technology to go to the stars at this time in this year, I would say yes! Absolutely! Why? Where's the evidence? That, my friend would not be mind-blowing. If you possessed the technology to travel through time and to distant planets in other galaxies, yet someone would use it to control you, or worse, harm you or your loved ones, would you let them have it? Probably not. This is one of the swords to defend your self with. Let it be known, technology is like an absolute power, it can corrupt! No matter how noble the intention, it can absolutely corrupt! Let us suppose there were a spaceship had a hard landing in New Mexico in 1947. If the occupants were in any kind of duress, they would have very well sent a beacon to call for help. The help being distantly related to the occupants would have been more careful. If I may use a simple scenario. If you were a field biologist in a relocation effort of an endangered species, you would tag and bag your target, and hope that you were not engaged by a concerned parent of that endangered species. And that parent, being as wise as it is, would probably stay back in the shadows until it is time to strike.

We very well may have captured aliens in 1947 and use items that we found from the site as our night vision, lasers, computers, and such, to help us in our daily lives. To make things for us a little bit easier. So the time we spend with our loved ones would be much more quality time spent. Though, I'm saying no more than the average person with average intelligence would say. I just choose to use a bit more of the fore brain than most try to. Telecommunication and non verbal suggestions to understand a person or persons in a set environment to navigate the landscape for your advantage.

If we now possess the technologies from aliens, we are already using it for our country's advantage. I do know for a fact that once you have something in your hands, it is only a matter of time before you break it, learn how to fix it, and then use it more wisely the next time. Such as our nuclear weapons. Now that we know the destruction and hardship it causes, we think twice about using it every time someone makes us angry. Now just imagine the harnessing of the blast of a nuclear weapon in a rocket aimed at Mars. All we'd have to do is, use small, yet controlled bursts of that energy to propel us toward our destination. Lead rods may do the trick. Just a suggestion. Just talk, don't know where I got that from.

We have been likened to children playing with matches. Some children have the intent to burn whatever they can, while some say hold on! Maybe that isn't such a good idea. And some may even say, let's use it for the good of all humanity. It is cold outside let us use it to start a fire to keep everyone warm instead. Thus everyone would sleep peacefully and a possible happily ever after. Not the case when it comes to certain inventions. Serbian-American Nikola Tesla (10 July 1856 – 7 January 1943) wanted everyone to have free electricity. Even an electric power plant to fuel our cars for thousands of miles. What happened there? Today, we still use his radio-coil invention from 1891. That should tell you how far we need to jump ahead.

Let me give you a look into the mind of an alien that wishes to deceive! Great! The earthlings went for the cellular bags of fake intelligence we have engineered and sent to them. They are proceeding just as we expected them to. They think they are actual beings that have been mated and created as they were. How silly of them! They seem to fall for anything they do not understand!

Let us rule their disobedient hearts and minds a while longer to give us time to really make them destroy one another, while we prepare the final weapon to annihilate them all. We can then rape them of all of that planet's resources, like we did Mars and many other planets then, once we get what we want, we can move on. They will then believe this to be their Judgment day! We will have a little fun while we are at it as well. We will abduct them because they still do not understand how we do it. We will study them and mate them with us and create a race that is truly worthy of yielding this technology. The new children will be with us, traveling through the galaxies of this vast universe to explore and pillage.

But we are not done yet! The fun, or should I say experimentation has just begun! Let us give them more time to marvel at what they still cannot understand. We will give them only what we want them to know at any given time to see if they can understand how to use true power. So, why give us this power in the first place right? There are several scenarios isn't there? One may be to see what we do with it and how we help each other with it. We are due for another jump in technology. There is also the scenario that if they give a select few what they want, and those select few operate with impunity (above the law), then aliens, in turn can get what they want. For example, abductions, bio-assimilation, whatever they want. I mean, if our government admit to something disarming nuclear weapons codes to our weapons, what haven't they admitted. To me, it's all in plain sight. Another page in another book!

Look, while there could be a thousand or more possible reasons why our government is doing what they are doing, what makes me giddy the most is that there has to be regular citizens with the real information, because ET's knew from the beginning that they could not trust people with power, i.e., our government! Smart! Because, neither did God. That's why he chose a poor kid, by the name of Jesus. Sure Moses was rich, but following a true God, you'd have to give that up in one form or another.

Aliens are by no means, ignorant. Look at what they have. I mean, we are still attempting with all of our bright minds to go to the moon again. Or to that red planet. Why do we kid ourselves the way we do? We have long had the technology to go to Mars and beyond. I tend to believe the people that the Government has tried to silence. The Bob Lazar, Buzz Aldrin and Gordon Cooper statements are what the mind tells us automatically is true. Yet, we have an opinion if it is or not. Why is that? Those are my true heroes. If I were Edward Snowden, I'd ask for a room in the White House and twenty-four seven protection. But I'm sure you know how that would play out. Someone would find a way to slip him poison and blame him for committing suicide on him self. We have some of the most gifted pilots on the planet. Yet, ask any one of them if they would start telling people about spaceships they often see when flying, and see what they say. Controlled!!! Fear, ladies and gentlemen is the thing that grips a nation. Even God said fear no one! Yet the guilty like to use it as a tool to force people to yield to their will. Sad days ahead!

I am not condoning leaking secrets but, I admire those that step out on a limb for all of humanity. If there is something that could benefit our humanity and not build us up to seeing hatred and human suffering then, by all means, release it. Please don't tell me what I am ready for! You would be surprised to know what people are capable of accepting and forgiving for once heard. Maybe even an apology. Just a Yes, we have it but we cannot disclose the full workings of it until we figure out how it works, so that if it falls into the wrong hands, we will not all be destroyed. How about that? I sure hope they don't take this book as an apology for their wrong doings! Now it seems like a Public Relations campaign just to keep it from imploding.

Imploding, hmmm! That's exactly what always happens to something or someone that can no longer understand how to contain itself while under pressure. Even our very earth cannot contain pressure once built up. It all needs a release! I don't like to stray too far off subject, big laugh… but, when officials seem to be on the side of the people, they always seem to get themselves assassinated! You already know the names. What is it about helping the world become a better place that just turns certain people off? At the rate that we can genetically modify seeds these days and making the seeds withstand any temperature. I find it hard to believe that starving nations cannot be abundantly overflowing with chickens and turkeys that are fat and adult in size in a matter of weeks to feed these people.

Now let us suppose that these beings are who they say they are. Just travelers, looking to seek out information about their own species. According to some words spoken by some person(s). That would make complete sense! Why not, right? They are far too advanced to let us try to figure out anything about them. Well, let's look at it from an intelligent alien's point of view. If I let you find my star map, and didn't wipe you out in the process, I already know with your history that I'd have a few thousand years before you'd even remotely figure out how to read it. So, why not give it to you. By the time you figure it out, you would have missed the chance to advance by another one hundred years or so. And the cycle of stupidity would just continue to go around and around and well, you get it. Now you have a shield to arm yourself with from the piercing arrows of evil while trying to spread the truth! Let us move on, because you know how I hate to waste time.

Now that we have this technology and are limiting it to very few people who are stressed out anyway. I mean just watch the news, right! We can start to see the bigger picture. I like to use a quote that Bill Gates used in one of his speeches somewhere while he was still getting rich. He said, "he would rather give a hard job to a lazy person, because that lazy person would find an easy way to do the job!" Well said, Mr. Bill Gates. Any successful nation has learned to share the load. Because taking from our own history and world history with expansionism and imperialism, how can you get richer if you do not allow the people to work! Allowing the people to share the technology, would ensure an equal and stable use of power and no just a wicked few would control.

Let's just say that all workers were conspiring to use everything they have learned at Area 51 (which does exist, because our wonderful government finally admitted it does) and decided to go rogue to Iran or Russia. Imagine how fast the government would act to not have such technology fall into hands it still has yet to trust. Now, if all the people in the U.S. as responsible and loving as we are, all had use of most of this technology, then we would be the eyes and ears of our country. It would be like our own national alert system that would immediately go into effect when someone tries to break in or intends to do us harm.

So, if I were an alien, which I'm not. Well, I don't know if I am. Then, why would I give you something that could be used to destroy me, unless I was using you to fire at those that were just getting in my way from taking over earth. I'd just fool the humans into doing what I want them to do, so I can continue to go about my business on your planet, doing what it is that I do best. Robbing the planet of all it's resources.

Chapter two

Beauty in C.H.P. uniforms

In 1996 in a small apartment complex, I lived with my now ex-wife and two children. It was at Via Hondonado Dr. in Penasquitos, San Diego, California. It was during the spring, as I remember going outside in a short sleeved shirt. This was during the time I was heavy into UFO's, (unidentified flying objects). I was doing research while watching a video with Stanton Friedman, Physicist and UFO speaker, when I saw a previously classified document on a V.H.S. tape. I stopped the tape to see what looked like an address and telephone number. It was to the Ministry of Defense, in Whitehall, London. I wrote the number down. Mr. Friedman then gave nine number/letters to previously classified documents contained at the Ministry of Defense. I wanted those documents.

The next morning, I proceeded to call the number and the ring tone sounded different. When the person picked up, I immediately explained who I was and that I wanted those particular documents sent to me. I was transferred once or twice. I heard the sirens of a police car that was different than the sirens we hear in the United States. For sure this was London, I thought. After the person asked me a few questions, I gave them the numbers to the files I wanted. I distinctly remember a series of three to four interrupting clicks in the telephone. I thought for a second. Could this be the phone tapping going on? I did not put it past them. By the way, this was the Ministry of Defense.

I immediately did not like where the communication was heading. What happened next blew my mind! It was the very next day, around 10 A.M. I heard the door bell ring. I looked through the peep hole and saw two white highway patrolmen standing at my front door. We were on the second floor. I opened the door and asked the men calmly, can I help you! With a blank stare a no movement of their bodies, the man in the front said in a mono toned voice, "you call us?" What the…I thought. I shook my head saying, no, I didn't call you! I turned to my wife at the time and asked her if she knew who would have called the police. She replied, no. It wasn't her and she did not know who called them. So, I looked back at the police and said we're sorry, it wasn't us! They just stood there, starring at me. I asked if there was anything else that I could do for them. They continued to stare at me. Now, it gets freaky!

I said to the C.H.P officers, "if there is nothing else, I will close the door now." They said nothing! Amazing! Super crazy, freaky! So, I slowly closed the door and immediately looked through the peep hole. They both were still there standing in the same position, still looking at my door. I said to my wife, this is weird. Then, about thirty seconds pass. The officers without speaking to one another, performed a simultaneous about face, and proceeded to walk down the steps to the patrol car. Yes, a patrol car. There was one minor detail though. They were CHP officers driving a San Diego Police Department vehicle. Isn't that illegal? What the hell? I thought!

Now, before I go chasing them, as I will and have my wife look out the balcony door to see if they come out, here's the other detail you may be interested in. The faces of these men were well, interesting. Yea, about that! They were beautiful! I mean it. So beautiful in fact, that there were no discernable pores to see. No facial hair. No lines from sun squints. Nothing at all! Zero! Imperfections did not exist on these men. The skin and the faces of these men had to have been prefabricated as they were. The closest I can come to describing the men is that they were plastic molded in high heat! After the face was molded it was somehow attached to the body. So, you feel I am making this up as I go right? Wrong! This is true. I am not the lying type. With this subject, I don't see how I can. But the pressing subject cannot wait! So, as I went to my kitchen window to look down at the car and the men, astoundingly they were still looking at me. While they were looking at me, the driver started the car, put it in reverse, and missed the garbage dumpster, by inches as if he has done it thousands of times. All of this while still watching me look at them in the window.

I told my wife get to the balcony now and tell me if you see them pass by. Within ten seconds, I was at the only entrance into the apartment complex. I walked up to the back of our porch and I see my wife there and asked if they came through. She said no. They did not pass her sight so, they must still be somewhere in the complex. No car can make it through the neighborhood and back to the street in under ten seconds. That would be impossible! But something happened. It was up to me to find out what. So, I did ask my wife to stay there while I searched the parking stalls to see if they went to another apartment. I searched and searched. Nada! They were no where to be found. The only way out besides the one way in was to drive through private fences and harming someone, and as I checked to see if that was the case. The case went cold! Right there! I literally stand and shake my head from side-to-side when I just don't understand something that should not be.

After carefully searching every stall, I proceeded back to the balcony, where my wife was still standing. I asked her if any car had passed by. And she confirmed, that there were no cars that passed by. None at all! I am not surprised to hear stories on the radio or on the news about strange incidences. I feel like a real life researcher. I feel like the actual involvement in these sorts of things, lend a higher amount of credibility to these accounts. Some say, once visited or contacted, you are continued to be contacted throughout your life. I don't mind. Because as long as they respect me, I will respect them! If they break that so-called line, I will not hold back my anger. And from the reports I hear. They can read thoughts, so they probably know what I'm talking about.

My first sighting was in Montgomery, Alabama. I was born and raised there! When I was around the age of fifteen, I witnessed my first space ship. This was really close! It was hovering silently above a transformer near the railroad tracks. It was the classic shiny, hamburger shaped disc with soft lights around the circumference of the ship. It wasn't a tiny ship. It was of some size. I often asked myself how something could make no sound with the mass it has! My second sighting was in La Jolla California. My wife, baby and I, around the 1992 time frame were enjoying the day watching the seals in La Jolla Cove. Also happen to be a favorite scuba diving site of mine. As we were watching the seals, our attention was turned to the sky. People were watching the planes fly overhead as they usually do. No pun intended! At about 1500 feet altitude flew over the last Cessna 152 it looked to be. It was flying east over the quay wall. That's when it happened! An Orange-glowing sphere came out of the western-facing water. It was quite bright- orange, and moving at approximately 30 to 45 miles per hour. I estimated it took at least 7 to 8 seconds to reach the altitude of the Cessna we last saw. And at that altitude, it's as if the sphere vanished into another dimension.

Let's get the missile and flare theories out of the way right now. My job in the Navy since 1985 to 2005 was an Aviation Ordnance man. I loaded bombs, missiles and rockets as well as flares onto F-14 fighter jets. I was well aware at the time that the smoke from a flare would have been evident, as it was very bright outside. Missiles fired from a submarine would also leave smoke. At the very least, it would leave water tails of some kind. And this sphere left none of those. It was a sphere that moved through water and vanished in thin air with absolute impunity. Although I am not a world renounced rocket scientist, I can not be fooled into believing that the lights over Phoenix, was the action of a flare. So, I can easily expel that notion! No matter what you are told about those lights over Phoenix, Arizona, they were not flares of any kind. Fact, flares smoke, and one would be able to see the smoke when it's the workings of a flare. And the smoke would linger. I would find out later in my Navy Reserve unit that even the helicopter agreed with the picture I drew him of the Silver disc I saw in Montgomery. He overheard me talking to some of my unit buddies about the incident, and he'd asked me to draw him a picture of what it was that I had seen. I drew the picture for him and he stared at it for a few seconds as I showed him the drawing! Wow! He said! He asked, "Is that is what you saw?" Yes Sir. That's what I saw, I told him!

You know, the pilot said. I was flying in Corpus Christi in my helicopter headed back to base and I saw this light headed towards me at a very high rate of speed. I thought it was going to crash into me, so I banked hard to let it fly under me. As I looked at it, before it went under me, it looked just like the picture you drew. What happened? I asked the pilot! It never came out from under me. It just disappeared under me. Like it went into another dimension or something, he said! My jaw just dropped! And everyone else that heard it eyes got bigger than a half dollar! They could not believe what they had heard from a Navy Helicopter pilot. But once again, those are my heroes. Here is to the Men and Women, who break their silence, on a subject that in one way or another, helps humanity! It is true. Without them, this world would be as boring as politics. The world is happy and somewhat reluctant to say they somehow, admire Edward Snowden, for what he did. But look at what it did. Now our President is apologizing to the German Chancellor for the NSA spying on her. I never recall anyone saying they had no clue it was happening! It is larger than we can expect. But for those that have a mind for the truth, will always find what we are seeking.

I have to tell the truth about what I see and hear. If I put it in my book, then I believe it to be true. We are told from an early age to shut our mouths, when speaking on a certain topic that someone a bit more powerful than us does not want us to speak about, isn't it? The same thing when we attempt to explain something we believe and that we know. Then, there is a certain atmosphere that erupts, with someone trying to explain what only you saw or heard. This is an attempt to dumb us down! To get us to believe what they want us to believe. To ""THEM, I say, "No more crap."" Let readers, read, and writers, write! That's what we do. We are the silent reporters. We write what our minds gave us the ability to write, and we do it without fear!

We can never seem to put a finger on exactly, who "THEY" are, can we? We may have knowledge on a few that pull the strings behind the curtain. But like the Wizard of OZ, they will be found out. There is quite the mystery, when it comes to who actually controls the publics on and off switch. That switch that most seem to have when abiding by the norm. But armed with more and more knowledge of the truth, and the anger of always being told how to think, is awakening the masses. Could it be the media? The deception of making us think we are getting the truth, when we are actually in a mental construct, like the Matrix!

You know the truth! We do not need anyone to tell us the truth. God himself spoke it in the bible. He said we have known the truth since we were children. "He said he put it in our hearts!" Today, the California Highway Patrol (CHP), tomorrow, the men in black, or white, or whatever! I just know, when embarking on the unknown, take some ammunition with you. Oh, and a camera. I've described thus far, the encounters that I have had. Now, what are yours? We all have stories to share. This is how I and many more people share theirs. And since the beginning of time, we learned by reading what has been written.

But for the sake of saving time, let's get to the next topic. When, we want to report something that seems out of the ordinary, who do we call? Ghostbusters, no! We call the police or fire station or FBI. These are the last people you want to call. So, Mr. Webster, what do I do? Okay! Keep a camera near by at all times. Video tape whatever the event is. Dictate by voice or in writing, the particulars by the minute of the event. If it is a Disc in the sky, state the color, the distance from you and the size it appears to be. State whether the movement was normal or wobbly. Write the date and time and place of the event and report it to Mufon. Hell, put it on Facebook and Youtube also. The more places you tell your story, the better the audience to understand what it is you saw. It's ok to report it to the police if you know them. Otherwise, be wary of the ones you give information to.

There's many tools at our fingertips to report Unidentified or Identified Flying Objects, (UFO's, IFO's). Mufon even has tools to help with determining if a photo is real or fake. You can also watch the most up to date UFO cases on the site as well. Is Mufon paying me to write this? Not a chance! I wrote this in the event you the reader, would like a fast tract to up to date information on UFO's. They (UFOS) come in all different shapes, and sizes. There are reports of UFOS the size of a basketball to the size of a city. And even continents. Yes, continents. Though, this is nothing new. The sightings have been around since time began it would seem.

From the time I was a child, I have seen things I had a hard time explaining. That doesn't mean I need to keep quiet about it though. Hey, like many of you, "I saw what I saw!" I will no longer let someone tell me I was seeing things! I already know I had seen things. That's my point! If you refuse to believe what it is that you actually saw, then you are easily influenced. And when you are easily influenced, you just may start to believe you did not see what you really did see.

It's not so uncommon, that many would like to make things up. Some would even like to string others along. Be wary of such individuals, because they may not be individuals at all! There are such beings on our very planet that shape shift, and morph into something out of this world. You and I have read most of those reports. You've heard guests on the George Noory coast to coast AM show, talk of kids with large, black eyes. They are looking for eat time, instead of dinner! They know English, but it is broken English. They seem to look like they have adapted to the dark all of their lives.

They seem to leave a trail of sickness and death wherever they settle, then as quickly as they came, they vanish. Still don't believe? You will, one day in the near future.

Chapter Three

The Odd, the Unexplained

In this chapter, I will just casually speak on odd and unexplained events, phenomena, and what was that! Wakey, wakey, no more sleepy, the Indians told me when building a Tee Pee. Even in old Indian stories, (some may say folklore), there are tales of sky gods and supreme beings that gave the Indians knowledge and a higher understanding, and wisdom. It baffles me that the understanding of man seems to have diminished, rather than keep up with time. There are so many tales of what goes bump in the night it would take a novel to write them all. Again, that's not my purpose. I shall arm you with words of wisdom to go on and not be fazed in the shenanigans of the tricksters.

If you listen to certain unbiased news stations, you would hear of spirits and cleansings being performed at an alarming pace. Catholic Priests get no sleep because of this stuff. But, if the Vatican start to announce just how many rights of exorcisms it has to grant to Priests, there would be an outcry. What is happening in our society, many will ask. The Vatican itself, stand on the grounds of so many secrets, that if they were all to be revealed, the world would literally stop! If the Vatican were to tell you that evil spirits are in fact Fallen Angels and there are countless numbers of them, you may not believe them. But, you may want to. If you believe in a higher being, then you must believe in the rest that goes along with it.

Fallen Angels, or demons, have walked the earth for a millennia. They work for the Devil. No secret there! They are here to take as many people with them as they can, because they know they have only a short time to do so. They hate men. Why? The reason is we are God's greatest creation along with his Son Jesus Christ. We know this though, right? So what's this got to do with oddities, Lawrence? Ahhh! You asked the right question. Demons have long been known to possess the power to transform them selves into almost anything with impunity. They do not have to believe in our physics or our science. They were part of the creation since the beginning. And being part of the first, there are certain perks with that job! And the Angels saw that man was a perfect creation. But wait, says the Angels! We are the perfect ones, God why have you taken pleasure in these pathetic creatures over us? And so, the Angels were cast out by the millions, banished from the heavens, forever.

Jealous little bastards! Now they have no Father! But wait, there's more. Since Lucifer was the head rebellious one, and stoked on his own beauty, he became the Father of the fallen ones. He was more powerful than the rest. He even wanted to overthrow God! Imagine that. The balls on that bird. Therefore, talking to a woman in the garden as a snake may not have had a big impact on Eve. After all, she is the weaker one. So, after Satan told his lies and deceived Eve, and Eve deceived Adam, we all must pay the price. Since Satan made the first two loose there way, we are on the same path. For we are the children of Adam and Eve.

Now, if morphing into a snake and talking doesn't get you, Satan also created hate in the garden. Cane killed his brother Able. Cane was very jealous of Able being God's favorite. But Cane, gave his leftovers to God as a sacrifice. Abel gave God his very best from his animals and from the food of the field. For this, Cane flew into a rage and killed Able. Abele's blood cried out from the ground. And so, we were cursed again to toil the soil all the days of our lives to work for our bread. Now, if you understand the power

Angels have, then you understand how they can form themselves into a child, or maybe even a bird, or a snake, and make us do and think of a lie as the truth, and the truth as a lie. So, if they have the power to turn into a mist, or a shadow, (after all, they learned how to do it since the beginning), how much more can they appear as someone you have known in the past. There are also good angels that have the same power. So, God always gives us ammunition to fight back with.

Demons are very smart. We are not as smart as they are. If we were, we would have figured out how to take a trip to heaven and come back as Satan does. He asked God if he could go and tempt his servant Job. God said yes, but gave Satan strict orders not to touch his life. Satan made homes cave in and killed Jobs children, put boils all over his body, and more. But Job, being faithful, kept his trust in what God can do for him. And as a result, Job was blessed many more time over. I know, I know this already Lawrence. I know you know. But did you know you can save on car insurance by switching to Gyco? Pun intended! Names changed to protect the guilty!

When asking about the things we do not understand, try asking ourselves about the things that we do understand. Like, how faith can move mountains. If that were true, then why not Fallen Angels morphing into animals or mist, and vanishing into a dimension we cannot understand. The power of prayer cannot be underestimated when performed. We can get things in life that we yarn for. We can call upon God and Jesus Christ to heal us when sick. But, it all has to be done in the way it is written for it to work. But, do not be deceived, Satan can grant wishes too. Be careful how you live your life. Never invite an evil spirit into your life. Stay away from Quija boards and the likes. No good can come of it. In times of trouble and in times of sorrow, always pray.

Now, that we have developed a case with the Fallen ones, we can clearly see why Aliens in the sense of the word, would not tell their secrets to man, (or at least, not all of them). Can you ask yourself that question? If you went to a new world to explore, and you had tools far more advanced than theirs, would you give a child matches? You know, the first thing children do is experiment with things that they know nothing about. So, before you go cutting into some fission reactor with a saw, try learning what may happen if you do! This is common sense right? Not so for those with their hands on technology unfound. Planes crash, weapons explodes, kills thousands, and still we frighten each other with the use of Nuclear weapons. Now, if I were an Alien that had an agenda with the humans, I might like that. You know, the plot to make each other destroy themselves with the most powerful weapons known to men. Then, as an Alien, I will have a clean conscious by saying, I only gave them the idea, they did the rest.

While something major is happening in the world, watch the stock market. The people go scurrying! Sell, sell! China is devaluing its own currency, for what? Uncertainty, yes but only for companies and investors. I recently asked my instructor in Corporate Finance, since the price of oil is so low, record lows in fact why has the price of gas slowly keeping up, and not falling with barrels of oil? He said because of there isn't enough reserves. So, if we don't have enough reserves, and the price per barrel is falling, how does that equal high prices at the gas station? Well, I'll let you in on a little secret. We have recently discovered that the earth has an abundant supply of crude. And it's not fossil fuel after all. It happens to be the lube of the earth itself. So, we have trillions and trillions of gallons of crude in the earth just waiting to be used. Literally, a lifetime supply of oil. That's why some politicians scream "drill, baby drill"!

Now, I could go on and on just writing things like this because, I know you like to read what I write! But, still I have to keep my word and keep it to the point. Now, about those skinny ass aliens, that's been lying to us. They suck! Major, man! I like the fact that they exist and all, but, they've got tricks all day up they're tiny little sleeves. If I ever get to shake one of their hands, I'll squeeze it hard to see if they change from gray to some other worldly color. For all I know, they are under the white house turning up some death ray machine with no one knowing. Those guys are tricky, man. And I mean "Really tricky".

Yea, they have all kinds of stuff. We got our night vision from them, lasers, and some even say Betty and Barney Hill saw them using a needle to extract fluid from Betty's stomach, which later turned into Amniocentesis. This is a process by which a medical professional can extract amniotic fluid from near a fetus to see if there could be any kind of abnormality that can possibly be treated before the baby is born.

The truth is, the intelligence these creatures or people possess is astounding. Their frontal lobe is so advanced that they can read our thoughts. There have been reports that they answer you using your voice. Yes, they cause you to speak the answer in which you were seeking. You think that's the only thing they can do? How about giving us the technology to blow out of the sky other supposedly threats by other aliens that they don't like. So, we are getting played in more ways than a football. We are not only on the verge of destroying each other we are killing other aliens as well. You know, as a recruiter in the Navy, this would be exactly the kind of influencing you would need to recruit thousands each year. It's like you're just sitting there collecting a paycheck and you're making the recruits do the paperwork for you. And you can go on about your merry way doing whatever it is you are busy at doing.

Well, maybe this is the case. As the counsel on reverse thinking, I propose that our over government department, (you know, those with the gigantic black budges), are experiencing a galactic episode! I will explain. If, indeed our government is in cahoots with our little foreign friends for technology, we can be in serious trouble. Get out with what you have while you can. Yea, I'm talking to you! When we finally realize that what we're doing may well violate our very own constitution! That part about foreign and domestic enemies. But, I guess as long as its kept secret, no one will know. Wrong answer! Everyone will know, soon enough.

I keep critically thinking about the part where we the people are not allowed to be part of what's going on. I guess the public cannot be trusted with new knowledge. I mean, someone really wants us to be in the dark. But why? The less we know, maybe the less we can help defend our country against hostile aliens, or better yet, alien terrorists. How can we help then? Report what you can. Use social media to upload any and everything so we can all be aware. Not left standing alone when the real war starts.

Credibility, is something that is given to another by the nature of status in a particular area. Such as, a pilot or an Astronaut. When Pilots tell or speak of spaceships moving at incredible speeds and performing acrobatics, in a way in which makes flying insects look like novices, I tend to listen. These are trained and intelligent men and women putting their lives at risk and told to shut up when they see something they were told not to talk about. Pilots and Astronauts careers are threatened by certain individuals if they say anything about objects in the sky. It is a shame that the very men and women we put into space and train so very well, when in the presence of a spaceship up there, they are immediately silenced. I like to call this, "from the cradle, to the grave syndrome," where one is born to do a certain job, but the particulars of the job go to the grave with him or her. Knowledge becomes power only when shared! When we all know our functions as a team, then we can say we have advanced. We are advanced beings. We just need to work harder at developing the things we need to develop to sustain our lives and health.

We often hear of the United States falling behind in Sciences and Math. If this trend keeps up at the rate that it is going, we will no longer have a foothold on the technology development we so desperately need to be productive on this planet. We have almost all the resources and then some, to start development on space travel. Redo some of Einstein's theories, and see them in a new light. The Tesla coil and other marvels, built for the good of humanity is being politically disassembled in front of our very eyes. Einstein showed us that we can see objects behind other objects by bending light, and Tesla showed us that we can have an abundant source of electricity by using a Tesla Coil to grab electricity from the air.

In the Navy, there are certain fields that require secret and top secret clearances. I, myself have a secret clearance for the knowledge of bomb building as an Aviation Ordnance man for the majority of my Navy career. I often talked to my departing Seaman and Airmen on their way out of the Navy or I would assist them in joining the Navy Reserves. I will not name names, (as if I could remember it anyway), so, I'll just say Seaman Smuckatelly, knew I was interested in UFO's and Aliens. He had no reason to lie to me, as I would have put him in his place. Being a recruiter, I could have did that. Well,

Seaman Smuckatelly asked me if I believe in UFO's. I told him, that it would be foolish to think we are it! He agreed. Candidly, he began to tell me how he was called to the Hangar Bay of a certain aircraft carrier and asked to try to find out how a certain electronic panel, (from the likes he had never seen before) worked. As he began his investigation, he was sworn to secrecy, as usual when alien technology can't be understood. He then began to poke and pry and tried his best to understand what it was he was looking at. Is it a Russian submarine panel? How could it be? There were no screws to be seen, and the panels were sort of built into the curvature of the panel. There weren't any type of assembly recognition of any kind. He could do nothing but try to expose some underlayment to see if he could see how it was made. The entire investigation yielded nothing. The entire time, he had no clue as to what he was looking at.

This is just a tidbit of information for you to be armed with, when embarking upon the unknown. Or, the known, but you keep quiet about it, syndrome. One thing about people that tell the truth is, I adore them. It takes a brave or foolish, heart however you want to look at it, to tell the truth! However bent it may be! Hence, the title. The truth is right there, we just have to want to see it. And often, it is hidden in plain sight. Our common sense tells us that our government is hiding something when it comes to extraterrestrial life here on earth. That is the way we were made to think. Though this is not a book on common sense, it sure wouldn't hurt to use it in our daily lives.

What's going on right now in Area 51? Well, let's take a journey of a different kind. This will lead us into our next chapter.

Chapter four

Area 51

Deep under ground lies an Alien base camp. Fortified with fighter jets and armed helicopters that buzz the sky over and near you if you get too close. Motion detectors of the likes in which we have never seen before heat sensors, to detect the body's temperature of a human or animal. Thermal sensors for night time security, advanced Flir (portable heat signature units), to capture any heat signature variation from the tumbleweeds and anything else around. Let's see, the aliens tell me I need to build a fortified bunker of the likes we have never seen before, that will withstand a Nuclear blast. It will cost billions of dollars and the politics surrounding it will show the public that there's nothing to worry about. Not until someone gets hurt.

The aliens want the humans to build them a bunker that reminds them of the pyramids so as to have a familiar place to regain their mental focus in the presence of humans. The humans are only worker bees. The aliens are actually controlling them. They have no clue. It is like a hypnotizing effect on the humans that work there. They go to work day after day with the only satisfaction being that of the pleasing of the divine ones that are posing as good little aliens. The aliens know that as soon as they get what they want, they will unleash hell on earth. They are giving the humans little by little bits of information that can be useful in a war-like situation. The bible states, that the first time the world was destroyed by water, and this time, it will be destroyed by fire! Statistically, more fires start in the desert than any where else.

They got us just where they want us! We jumped out of the pot, and into the frying pan. At The Little Alien Inn is buzzing with chatter about what, when, how, and why saucers go unchecked through the skies above the little town of Rachel, Nevada. No plane we have can catch them. They are immune to capture. Unless the unlucky ones being shot out of the sky with a new technology from the aliens. What they really did was give the humans access to one of there weaker weapons system. Because why would they give us the more powerful weapon, that is more sophisticated? They wouldn't. Under ground, they give commands and humans follow. The humans are allowed to use certain tactics, but the aliens are always one step ahead. They read minds through brick walls too.

No one just stumbles upon one of the greatest finds in the history of man kind. It primarily fell right in front of us, and pretended to be crippled. Their death is not like our death. They can return and regroup, because they also know the secret of life and death. Their years are longer than our years do to the fact that they were not cursed in the garden with the years that number about 120. Do you really think they will just give up without a fight? They will just freely give up thousands of years of technology to a race of beings that still can't settle their differences with one another? Someone seriously, failed. I wonder who dropped the ball. This new technology comes with a price. It's not like there's no biblical relation of this. God said that, in the latter times, knowledge will increase! So, it's no secret that, in these latter times we are smart. We are smarter than we were years ago. It's just we really didn't advance until the saucer crashed in the desert and propelled us into the next century. Things happen for a reason, yes. But the reason, I fear, is not our own reason. The reasoning of some thing, the aliens! The reason of treason to take control of a select group of people that will yield to them. I like to call it that because in the end, the government will be overthrown. Chaos will come upon the streets. The people in the cities will starve and have no where to go. The people out of the cities will be sick, and have diseases. The reason is doom!

While the humans are basking in their glory, the countdown has begun. The aliens will want nothing extravagant, as their human counterparts do. The humans want fame, notoriety, and money. These things, from the beginning, have corrupted man to his very soul. As an alien, would I not want you to be this way? I would want you to have all that you desire. But what about the things you are forgetting to ask for. You ask for weapons and the ability to conquer your adversaries. Why not? Such a barbaric want! As an alien, I would give it to you. All to keep you in the dark about what my true mission is.

About that! To learn your inner most desires and give them to you. Then, you will be under the alien's command. Think about it. Can you take a vacation at all? Or do you work with unknown substances while the aliens to not know what you talk about. Those unknown substances are causing cancers that cannot be cured by mans understanding. Because it's the start of the disease, that will be unleashed on the people in the cities. It will be an apocalypse. A contagion that's uncontrollable. The aliens know what the cure is, it is very simple to them, but they say they don't know, they're lying. While the weapons they gave us so far, are preventing the good aliens from coming to earth are working, the good ones are on the realm of the good angels. They will fight together to unleash punishment upon man.

In all my days on this earth, I never thought that I would be writing this short book. But lately, I could not help myself. I needed to write this. This is just the bent truth.

Is it real? Of course it is. I will not lie. Especially about this! This is already a touchy subject as it is. It would not benefit me at all to make things up. I seriously believe that aliens are literally controlling humans not just at Area 51 but in many other places throughout the world, and in space. Have you not heard of secret space missions before? Those are the people that never seem to come back home. Some do not even have families or, any family that care for them. But for the most part, they are everyday Joe Smoes that seem to be involved with the program.

Any day, one can see the planes that take off from the Las Vegas Airport with blacked out windows. I see, so you can't even see outside! How controlled is that? From the very moment the workers get on the plane, they are controlled. There are even reports that the workers get telephone calls early in the morning reminding them of their duty to return to the plane each day. I doubt if one of them even call in sick. Imagine the trouble one can get in if they call in sick! There comes a surveillance team to control the matter. Because, as a protocol, (sick or not sick) is nip it in the bud! Spying on your worker day! That's another term I'd like to coin!

So, your going to give me a good paying job that's super secret and I can't give anyone a hint as to what I do? Ok, sure, why not? I'll take it. I've got two words for you, Edward Snowden, you know him. So, if this stuff is so out of this world, why haven't anyone came out and said anything? Hold on Nellie! Bob Lazar. He worked in S-4 Groom Lake and back engineered flying saucers. He was a Physicist that did not understand the workings of the ship. Though, sometime later, he explained publicly what he was doing at Area 51. He went on television and told people the truth about what was going on there! Ever so often, mind control does not work on individuals with strong wills.

Speaking of Bob Lazar, he has since started his own company called United Nuclear, and also helps the aircraft makers in very important regarding their aircraft structures. Wait! Did he not work on reverse engineering space craft in Area 51? Did they not try to erase his past, as if he were some nut case, Wanna be? Then why on God's, green earth, would companies hire him to do such things unless he was really telling the truth about what he did at S-4? It makes a lot of sense now. Hire the ones you let live, because they have a special other worldly knowledge now. According to Bob, time travel was already possible when he first came out and told the public how it worked. Bob Lazar, in an interview, explained how the ship was like a piece of paper lying flat on the table, and one side represents a place in time you want to go to, and the other side represents the side that you are on. Now, to get through that fabric in time, you would have to pull that point in time to you and you would make the jump. So, the pictures and videos of space craft seeming to go faster than our eyes can keep up with are actually jumping to that spot in which they stopped at.

Nowadays, Bob Lazar will not even talk about spacecraft or alien technology. And he now consults with some of the very ones that said he never worked with them. I wonder if the companies he consults with now might do the same thing to him again in the future as they did to him in the past! I could tell you where Area 51 is and the longitude and latitude on the map. But, I want you to find it, it's in Nevada. Use Google Maps to get a real (maybe), bird's eye view of the place. You will be amazed. The government used to deny all the time that the base did not exist. But, now there is just too much overwhelming evidence to the contrary. Right now, our national debt lies around 18 trillion dollars. How much of that you supposed went to black projects? I'd say half of the 18 trillion dollars up until now, secretly went to the black projects. Look around you. How many man hours were spent on developing all devices and things we see and hear? Some things we hear but still can't see. Like the trumpets going off all around the earth and the loud explosions that do not register on the Richter scale as an earthquake. What is that all about?

As humans, we love to pretend. We should not continue to progress in this manner. The signs are very clear to many. The reality is that some would like to keep it that way. They are scared to be called crazy, even when they are leaning in that direction. Sure, you will be fine as long as you take these pills. Isn't that what you always get, when you go see a doctor? I know it is true because I am a disabled veteran. I frequent the hospitals a lot. But I also ask the providers not to prescribe me more medications. You have to be very careful these days. Let's be honest, we do not all read the side effects as we should. Some meds give you more symptoms of other illnesses, while trying to cure a current illness. What!

You may be saying, why's Lawrence talking about meds. As a matter of fact, meds are also used in mind control. In some accounts, many people have come forward and explained how they were drugged after being showed alien technology, as to discourage their brains from divulging what they had seen. You can google it, at google.com. Certain shows will pop up. Watch them! Be careful, and be watchful! Sometimes, the control can be as subtle as a new person entering your life. Ever been separated or divorced and now living with the one that you thought you'd never be with and ended up just calling it fate? Is it really fate? Or maybe it's just another form of control on a chemical level.

One can never be too careful these days. We have grown leaps and bounds over our wildest dreams since the 1940's. We have technology, that while when used may not be admissible in court, can definitely improve Law Enforcement's criminal apprehension techniques. We have laser pointers and high speed computers and 40 megapixel phone cameras that were unheard of just a few years back. Just imagine it, and we probably already have it. If you think we don't have it, search for it on the internet. Just google it! I can give you a few scenarios that we haven't came up with though. What about a device that when used, would convert any animal's sounds into a language we can understand? What about a scanner that tells you when you have any form of cancer cell in your body, and immediately gives you a food source to consume to eliminate the cancer cell (s)? I doubt that Area 51 would release either one of those devices to the public.

Imagine that aliens do not get cancer, and if they did, it would probably cause them to live longer, than to die sooner! It would also be conceivable, that they would not like to give that information over to us because, face it, they want us to die. Not just to die only, but to do work for them as there technological slaves only to propel their agenda further. We get some slowly developed piece of technology to use over other leaders in this world, to continue the thought that we are the super power still. And that may well be! But it does not stop the (what I call) the "Moon visit Theory" people still believe! I mean, sure, we are a strong nation. But look at our debt and debt to other countries that out populate us. We have around three hundred million people in the United States. Look at China! Look at India! So, our only power is not in the people, as it has been talked up to be. It's in our weapons mass of destruction. So, who sold that information to other countries anyway, on how to make a nuclear bomb? And who sold it to Iran?

You've read and saw the videos on the science channel and the discovery channel about how spaceships seem to dart in and out of places like nuclear sites with impunity. And why not, right? If you could move at the speed of light, what would you do? Hit up the closest bank? Take a few Big Bite hotdogs from 7-eleven without paying for them? Don't forget to put that nacho cheese on it! Everyone can speculate what they could or would do, if they possessed certain abilities. That is the notion that keeps our spy agencies one step ahead, on some issues. Notice I said, "Some" issues? We sure don't know enough to keep all secrets safe. No matter how much we try. But just to let you know, your secret is safe with me! Let's pan the camera back over to Area 51, for a moment shall we. Time travel and super fast speed has to be at the forefront of at least one of their projects. I mean, why go through the hassle of building a larger run way right. And still, even the Chinese know Area 51 exists. And so does the whole world.

So, you're driving down a dirt road in Rachel Nevada and you come up on this sign that says, WARNING deadly force authorized beyond this point. Okay, that would get me more interested in knowing what's behind those invisible fences. Then, as we turn our attention to the hills, a truck with people in it slowly moving your way, as you decide to test the sign made of wood. The Men In Truck, knows you're scared because you read the sign. I'm just going go jogging past it one day and tell them I think I lost my drone, and ask them if they can take me for a ride near the S-4 hangars to look for it. I'll tell them it's got live feed to Youtube.com and an extra long battery life to scare their asses back. All it takes (if it hasn't happened already) is for one good hacker to get in there and do the world a lot of good. This is going to be a day we all long for.

Why will a hacker get into Area 51? Well, the answer is simple. To break the code would be a heaven sent. That will be the straw that broke the camel's back! All the sound and camera feed will be recorded and sold to the highest bidder in some other country. Like I said, this could be already going down. I know they run a tight ship. But I have one question for anyone that works at Area 51. How long do you think this is going to last? Think about it. Everyone, who is someone, wants to know the full goings on in Area 51. From aliens, to time travel, we want to know it all! What I could not figure out though, is why the aliens wouldn't want the world to know about time travel. Probably like them, we'd all run the hell away from this get up! I mean, look around, there's got to be some cute, purple guys and girls out there some where right? But if you get caught, don't say I didn't warn you. Think about all those abductions around the world. The hybrid children that has been reported and the cattle mutilations reported by Linda M. Howe.

The way I see it, is Area 51 and all of its occupants are going to have a lot of explaining to do. There are no words to explain it, when the curtains come falling down. You think the iron curtain was tough! I can think of one movie that pretty much sum the aftermath of it all, Independence Day, with Will Smith. Remember that? The guy to the President, "what about Roswell, You knew then, and you did nothing!" Then, the grand tour will happen. Wow! I can see it now, television cameras, news crews, helicopters, loads of money to get the first person to talk to the public about what's going on inside S-4 at Area 51. World-wide teams of people unlike the numbers no one has seen before.

Every flight will be booked. Airports will be filled to capacity. People will buy and sell tickets to Nevada as if they were scalpers selling tickets to a concert or basketball game.

If and when news drops about Area 51 being opened up, it will be equivalent to news about the atom bomb. It will be a moment of reckoning. I'm sure George Noory will be there and I can't see why I won't be as well, unless of course I'm in church. Nothing takes my attention away from worship. That's the one thing in this world that keeps me sane. All the UFO buffs from around the world, (that can afford it) will be there. If not, they will be the unsung heroes of their towns, cities, and states.

Though, no amount of coverage will make up for the billions of dollars of taxpayer monies that went into that place and more like it. For what reason is that justifiable? Just to be in cahoots with aliens? No matter how you look at it, we are not given the fare share of what they know to better our lives. We can be light years ahead of our time. We can beat the diseases that cause so many deaths around the world. According to Reuters 2015, at least 11,000 people have died in West Africa since the deadly outbreak 18 months ago. Where's that Alien technology we need now? The time lapse since the Roswell UFO was shot down has afforded us (undoubtedly) various payouts in culture advancement. Though, it seems we only want, or focus on the destructive part of the technology instead of the truly important part of it.

I cannot fathom that we are sitting by blindly being controlled like shepherded cattle being led to the slaughter! We are more than that. We were not put here to be fed to anyone or anything. We are God's greatest creation. That's why Satan hates us so much. Amazing, huh? We have God on our side, though he's not going to come down and say to the few select, "be nice to the masses now, play fair!" That's up to us. Remember, God helps those who help themselves! Unfortunately, we do not live in a perfect world. We keep trying though. There may be some good that come out of Area 51, just not now! Imagine, living in a new world that just came out of a primitive state just centuries ago. And here we are, with a hidden government or two, aliens on a secured base that even the President refuses to go to, and all it has gotten us is, Nada!

Aliens are not going to let us have the matches. They are very smart. Because they know the first thing we would do is kill them, send a Nuke to there home planet, then use the weapons for the wrong purpose. Why would you keep secret, what you plan to do for humanity, if it were good? If I know that, sure E.T. knows that. They know, in our primitive thinking, we cannot totally be trusted. There is a reason they can read minds. This is so, that they could stay 10 paces ahead of us. If we could even trust one another, they would know that. From what I recall, they were not the first ones that shot. We have a knack for shooting first then asking questions.

The shooting first method can work in some instances. I wonder, if when we shot down the first UFO we knew what we were getting ourselves in to. Apparently we have no idea. If we only knew that the whole idea of getting a jump start on beings that were so far ahead of us, we would have attempted to make peace. Not hold them hostage. But, If, I were an Alien, I definitely would make them think they were holding me hostage. I mean, I got time to loose anyway. I don't age like you do. I live for hundreds of years. What's twenty or thirty? Just enough time to watch my plan come together. I'm just saying, if you have no respect, even for the thing that is smarter than you, soon that thing will destroy you! Even time it self, soon comes around.

There are numerous people that have paved the way for us to believe the unbelievable. The most important of these I believe, was Bob Lazar. Even though I have had a fascination with the silver discs way before Bob came on the scene, he was the one that told what he had done. He told it in a way that when investigated, a cloak of secrecy developed on the part of the people he claimed to have worked for. If there are those that try to close doors so others can not achieve greatness, there will be more doors open. One thing they could not do is take away his knowledge. I do believe the Aliens in Area 51 liked Bob and knew what he was doing. For the fact that he was back engineering the craft, tells me that the aliens did not want us to know. This is also part of my Bent Truth theory to support my belief that aliens have an agenda that we are too closed-minded to understand.

The truth is something that hurts many people when told. Offense is taken to protect the offender's ego, not heart. Because the heart is always willing to tell the truth, many want to but are told that there are consequences if they do. So I say this to the brave and open-minded, never give in the something that will ultimately lead to your demise. The only one you should fear is God himself. And God is not in Area 51! Sorry for any believers that feel that they are doing something right, just because it pays well. But think of your real future. Do you really have one if you reveal what you really know about your job? Well, Bob Lazar has a great future, and he is getting richer by the day.

Conclusion

This short book, (if you can call it a book) is the other side of reality. What's going on in the UFO world is deception on the largest scale. Some may even call it a cosmic water gate. Unlike just aiming missiles at our enemies, we now aim them at ourselves and everyone else in the world. If those weapon systems are hacked into, forget about it, just pray. So join the many that have embarked on a mission to expose the truth, and let's invent our own technological advances without relying on aliens or evil Angels to influence us or keep us from being great! And in the end, try not to be consumed by **"The Bent Truth!!!"**

Rest in Peace, Boyd Bushman!

References

Youtube.com Zeta Reticuli Aliens: Robert Lazar 1 of 4

Facebook.com

google.com

http://www.mufon.com/research.html

http://mozartsrucken.blogspot.com/2012/02/upgrades.html

https://www.youtube.com/watch?v=oy96TvyxDdY

www.ingramcontent.com/pod-product-compliance
Lightning Source LLC
Chambersburg PA
CBHW080651180526
45168CB00008B/3387